I0402516

SUDOKU

100 PUZZLES WITH SOLUTIONS

MULTI LEVEL **BOOK 1**

ISBN: 9781694062819
Copyright © 2019 Tim Bird.

A standard Sudoku puzzle consists of a grid of 9 blocks. Each block contains 9 boxes arranged in 3 rows and 3 columns.

Column   Block   Box

8	4	1	7	9	6		3	2
			2	5	4	8	1	
6	2	5	3	1		9	4	7
2		8		3	7	4	5	1
3		7	4	8		2	6	9
	5	6	9		1			8
5							2	3
1	6	3	8	4	2	7	9	5
9	7	2	5	6	3	1	8	4

— Row

The Basic Rules of Sudoku:

- There's only one solution to a Sudoku puzzle. A puzzle is considered solved when all 81 boxes contain numbers by following the Sudoku rules.
- When you start a game of Sudoku some blocks already have numbers (fewer number make a harder puzzle). These numbers cannot be changed.
- Each column must contain every number from 1 to 9 and no two numbers in the same column can be duplicated.
- Each row must contain every number from 1 to 9 and no two numbers in the same row can be duplicated.
- Each block must contain every number from 1 to 9 and no two numbers in the same block can be duplicated.

Here's the solved puzzle:

8	4	1	7	9	6	5	3	2
7	3	9	2	5	4	8	1	6
6	2	5	3	1	8	9	4	7
2	9	8	6	3	7	4	5	1
3	1	7	4	8	5	2	6	9
4	5	6	9	2	1	3	7	8
5	8	4	1	7	9	6	2	3
1	6	3	8	4	2	7	9	5
9	7	2	5	6	3	1	8	4

Puzzle 1
Easy

7	8	3	1					6
	6		7	5		3		2
	4	5	8			9	1	7
	1	4	9	6			2	3
	2	8	3	7	1			9
9	3	7	2	4	5	1	6	8
4	7	2	6	1			9	
8	5		4	9	7	2	3	1
3	9	1	5	8	2	6	7	4

Puzzle 2
Easy

1	4	6				9		7
5	9	2	6	7	4	8	1	3
3	8	7	9				6	4
4	5	1	2	6	9	3	7	
2	3	8	1	5	7	4	9	6
7	6	9		4		1		
6	2	5	4			7	8	9
9	7	3	5	8	2	6	4	1
		4	7	9	6		3	

Puzzle 3
Easy

8	6	5	1		4	3		
9	7	1	3	6	2	4	8	5
2	4	3	5		8	1		
4	5	7	8	2	3	6	9	1
3	8	2	9	1	6			
6	1	9	7	4	5	2	3	8
7	9		2	3	1			
1	3		6	5	9			
5	2	6	4	8	7	9	1	3

Puzzle 4
Easy

1		8		2	4	7	9	3
9	7	2	1	3	8			6
	4	3	7		9	2	1	8
8		9	2	4	7	1		5
	2	1				9	8	7
		7		9				2
2	1	6	4	8	5	3	7	9
3	8	4	9		2			1
7	9	5	3	1	6	8	2	4

Puzzle 5
Easy

4	8		1	5	2		7	3
9	5	7	4	3	8	1		2
2	1	3	9		7	5	4	
1			8	7	4		5	6
5		8	3	2		7	1	4
		4	6		5	8	2	9
	2	1	5		3	4	9	7
	4		7	9		2	3	1
		9	2	4	1	6	8	5

Puzzle 6
Easy

		4	5	8	7	9		6
		5			3			8
7					1	3		5
9	2	7	4	5	6	8	3	
4	1	3	8	7	9	5	6	2
5			1	3	2	4	9	7
8	7	2	6	9	4	1	5	3
3	4	1	7	2	5	6	8	
6	5	9	3	1	8			4

Puzzle 7
Easy

	6	8	1	2	4	7	5	
	7	5	6	8	3			
2	1		9	5	7	8	6	
8	3	1	4	6	2	9	7	5
6	5	9	3	7	8	4	2	1
7	4	2	5	1	9	3	8	6
	9	7	2		6	5		8
	8	6	7		5		9	
5	2		8	9	1	6		7

Puzzle 8
Easy

			6	4		3	7	9
		4	9	7		6	5	2
	6	7	3		5	8	1	4
7	8	3	2	5	6	4	9	1
6	4	5	1	8	9	7	2	
2	9	1	7	3	4		6	8
			4		7		3	6
4			5	6	3		8	7
3	7	6	8		2		4	5

Puzzle 9
Easy

9	3	7			2	4	5	8
6	8	4	5	9		3	2	
	2	1	4		8	6	7	9
1	4	3	8	7	9	2		
7		8	2	5	6	1	4	3
			1	4		9	8	7
3				2	5	8		
8	1	5	9	6	4	7	3	2
4	7	2	3	8	1	5	9	6

Puzzle 10
Easy

	2	9	1	8	7	3	4	5
3		7			6	8		9
4	8	1		5	3	7	6	2
2	9	4	7	6	1	5	8	3
5	7	6	2	3	8	1	9	4
8			5		4	2	7	6
9		2	8	7	5			1
7	4	5	3	1		6	2	
1	3					9	5	7

Puzzle 11
Easy

4	3	5	8	2	9	1	6	7
8	6	2	3	1	7	5	9	4
		9	6	4	5	8	3	2
	5	8	2	3				9
9		3	5	7			8	1
			9	8		3		5
			4	5	8	9	1	3
5	9	4	1	6	3			8
3	8	1	7	9	2	4	5	6

Puzzle 12
Easy

6	9	3	8	1	4	2	7	5
5	2	8	6	7	9	3		
7	4	1	5	2	3	6	9	8
2	5	6			8		3	
4	3	7		5	2		8	6
8	1	9	3	6	7	4	5	2
		5	2					
		2	4		1	5	6	
1	6	4	7	9	5	8	2	3

Puzzle 13
Easy

7	1	8	9	4				
4	2		3	1			7	8
3	5		8	7		1	4	
1	8	7	2	9	4			6
5	3	2	6	8	7	4	9	1
9	6	4	5	3	1	2	8	7
2	4	5	7	6		8	1	
6	9	3	1	5		7	2	4
8	7	1	4	2				

Puzzle 14
Easy

1		7	6	8	4	2	5	9
8			1	9	5	4	7	3
9	4	5	3	2		6		
		9	2	3	6	7		8
7		8	5	4	9	3	6	
2		3		1	8	9	4	
6	8		9		3	5	2	4
3	7	2	4	5	1	8	9	6
5	9	4	8				3	7

Puzzle 15
Easy

3	6	7	8	1	5	4	2	9
5	8	2	4	7	9	1	6	3
1		4		2	6	7	8	5
6	5	3	1	9	7		4	
				5	4		3	
				8	3		5	
2	7	5	9	6	8	3	1	4
9	3	6	5	4	1		7	
			7	3	2	5	9	6

Puzzle 16
Easy

	1	9	8	5	4		7	2
5			1	2	7		8	9
7	2	8	6	3	9			
9	7			8	3	2		
8	5	2	9	1	6	7	4	3
1				7	2		9	
4	8	7	2	6	1	9	3	5
2	9	5	3	4	8	1	6	7
		1	7	9	5		2	

Puzzle 17
Easy

	1		9	5	6	4	3	7
			2	4	1	6	8	5
5	6		7	3	8	2	1	9
8	9	2	1	6	7	3	5	4
6	5	1	3	8	4	7	9	2
7	4		5		9	1	6	8
4	2	6	8	9	3			
			4	1	5			6
	8	5	6	7	2	9	4	3

Puzzle 18
Easy

6						7	9	3
9	7	4	6		3		2	
8		3	2	7	9			6
3	8	7	1	9	5	2	6	4
2			7	8	4	9	3	5
4	9	5	3	2	6	1	8	7
7	3	9	8	4	1	6	5	2
1	4	8	5	6	2	3	7	9
5			9	3	7			

Puzzle 19
Easy

4	6			5	8	7	2	9
9	8	5	1		2	6	3	4
7	2	3					1	
5	7		6	2	9	3	4	1
6	1	4	5	8	3	2	9	7
3	9	2	7	4	1	8	5	6
1	4		8	3	5	9		2
8	3	9	2	1		4		5
2	5					1	8	3

Puzzle 20
Easy

	2	3		4	8			5
8	4			3	5			2
1	5	9	7	6	2	4	8	3
2	7	1	4	5	6	8	3	9
4	6	8	3	7	9	2	5	1
3	9	5	2	8	1	6	7	4
5	8	2	6	9	4	3	1	7
	3	4		1	7		2	
	1			2	3		4	

Puzzle 21
Easy

6							9	
9							6	
8		3		6	9	7	2	4
1	8	5	6	9	2	4	3	7
2	3	9	8	7	4	5	1	6
4	6	7	3	1		2	8	9
7	2	6	4	3	1	9	5	8
3	9	1				6	4	2
5	4	8	9	2	6	1	7	3

Puzzle 22
Easy

4		2	6	1	8	3	7	
		6	7	2	4	5	8	1
7	8	1			5	2	6	
2	4		3	6		8	1	7
6	7	8	2	4	1	9	5	3
1	9	3	8	5	7	6	4	2
	2	7	4	9		1		
5	6	9			3	4		
3			5	8	2	7	9	6

Puzzle 23
Easy

	6		1	9	5	2	8	3
1	3	8	7	4		9	6	5
9	5	2		3			7	4
5	2	9	4	8	6	7		1
	4	6		1	9		5	2
8	1		5			4		6
2	9	1	8		3		4	
6	7	4		5	1	3	2	8
	8	5		7	4	6	1	

Puzzle 24
Easy

7	1	4	6	9	2	3	5	8
	8	9		5			2	1
	5	2	1		8			4
2	7	8	3		1	9	4	5
		5	9	8	7	2	1	3
9	3	1	2	4	5		6	7
	4	7	8	2	9		3	6
	2	6					8	9
8	9	3			6	4	7	2

Puzzle 25
Easy

1	8	7	4	3	9	6	5	2
			7	6	1	3	9	8
9	6	3	5	2	8	7	1	4
5	4	8	9	7	6	1	2	3
6	3	2	1	4	5	9	8	7
7	1	9	3	8	2	5	4	6
	9			1	7		3	5
3			8	9	4	2		1
					3			9

Puzzle 26
Medium

4	8	7	2	1	5	9	3	6
1	6	3	7	9	4	5	8	2
2	5	9			8	7	4	1
7	4	1						
5	9	6		8				
3	2							
8	7	2		4		3	1	5
6	3	5	8	2	1	4	9	7
9	1	4						8

Puzzle 27
Medium

			3		8		9	2
2	6	4	9	5	7	1		
			2	1	6	5		7
	8	9	5	6	1	7	2	4
			4	3	2	8		9
4	2	6	8	7	9	3	1	5
6		2	1	9		4		8
			6	8	4	2	3	1
1	4	8	7	2				

Puzzle 28
Medium

4	2	3			8	9	6	7
	9	1	3		7	2		4
	6	8	9		2	5		
3			6	7	1		2	5
	7		5	9	4			
			2	8	3	7	9	
8		6	7	1	9		4	2
	1	4			5	6	7	
9	3	7		2	6		5	

Puzzle 29
Medium

		5	1	3	8			6
		1			5			3
6		3		9	7		5	
2	9			1	6		3	4
	1	8		4	9	6		7
		4			2	1	8	9
4	5		6	8	3	7	1	2
1		2	7	5	4	9	6	8
8	7	6	9	2	1	3	4	5

Puzzle 30
Medium

7	3	6			2	9	8	
9	2	1				5	3	4
5	4	8				7	2	6
3	8	4		2		6		5
2	6				5	1		3
1	5					8		2
4	7	5				2	6	
8	9	2			4	3	1	7
6	1	3	2			4	5	

Puzzle 31
Medium

		9	4	7	1	5	6	3
7	5	4	9	3	6	1	2	8
6	3	1		5		9	7	4
9						4	5	
1		6	5	4		3	8	
5	4					7	1	
4	9				5	2		1
3	6				4	8	9	7
8		2	7	9	3		4	5

Puzzle 32
Medium

3					8	2		5
8						4		3
2			3		4	8	7	6
1	9	3	8	7	5	6	2	4
7			1	3	2	9	5	8
5	8	2	4	6	9	7	3	1
4				8			6	9
6	3	5	9	4	7	1	8	2
9		8					4	7

Puzzle 33
Medium

	2	5	4	9	1			
	4	8	5	6	2		9	1
6	1		8	7	3	4	5	
4	6		1	5	9	8		3
9	5	3	6	8	7	2	1	4
8		1		2	4	9		
	3		9	4		1		
		7		3		5		
2	8	4			5	6	3	9

Puzzle 34
Medium

5		4	8	1	3	9		
	8	1			2	7	4	
	2		4		7		8	1
	9	8		4				
	5	7		6	8	1	9	
	1		5	7	9	8		
1		2	9	8	5	4		7
8	4	5	7	3	6	2	1	9
		9	1	2	4	6	5	8

Puzzle 35
Medium

2	3	6	1	9		5	8	4
4	7	9	3	5	8		2	
8		5	4	2	6	3	9	7
5			2	4	3		6	
9			8	6				
3	6		9	7				8
6			5	1	9	8	7	3
7	5	3	6	8			1	
1	9	8	7	3				

Puzzle 36
Medium

4	5	9	1	6	8	3	7	2
8	1	3	2	7	4	9	6	5
6	7	2			3		8	
		1		3	7			6
7		6					1	3
	3	5	6		1			
3	9		7	1	5	6		8
5	6		8					
1	2	8	3	4	6	7	5	9

Puzzle 37
Medium

1	8	7	5	9	4			
2	9	6	8	3	7	1	4	5
5	4	3		6	2		8	7
	7							
				7	6		5	
4	6	5	3	2		7		
9			2	8	1		7	
7	2	8	6	5	3	4	1	9
6	3	1	7	4	9	5	2	8

Puzzle 38
Medium

6	4	3			8	7	9	
9	8		4	6	2		5	1
5	2		7	9	3		4	
1	3			2				
	7	2	8		6		3	9
8	5	6		7	9	1		4
2		5	6	8	7	4	1	
		8	2		1		7	5
7		4			5	2		

Puzzle 39
Medium

9	3		7	4	5		6	
	4		9	1	6		5	
7	5	6	2	3	8	9	1	
	9	4	5	8	2			
2	6	3	4	9				
5	8	7				4	2	9
6	2	9	8	7	3	5	4	1
3	7		1		4		9	
4	1		6		9		7	

Puzzle 40
Medium

5				8	9	6		
9	2		6	3	5		7	8
	3	8		7	2			5
7	6	2		4		8	5	1
8	9			1	6		4	2
3				2	8	7		
	7	9	3	5	4			
4	8		2	9			3	7
2	5	3	8	6	7		1	4

Puzzle 41
Medium

5	3	8	9	7	2	6	1	4
6	1	2	4	3	8	7		
9	4	7	5	6	1		8	
							6	
		3		2		1		
	6			5		9		
1	8	5	2	9	7	4	3	6
3	7	9				5	2	1
4	2	6	3	1	5	8		

Puzzle 42
Medium

4	5	1			2	3		
	8	2		3		9	4	5
	3	9	8	5	4		1	
3		4	2	7	5			
5		7	3	8			2	
2	6	8	4	1	9	5	7	3
1	7	3	5	6	8		9	
9	4	5		2		8	3	6
	2	6	9		3		5	

Puzzle 43
Medium

	9		1					
7	3		9	5	8	4		
	5		6				8	
5	4	8	2	3	9	6		
3	1	7	5	6	4	2	9	8
6	2	9	8	1	7	5	3	4
4	7		3	8		9		
2	8	3	4	9	6	1		
		5	7	2	1	8		3

Puzzle 44
Medium

6	3	2	5	7		8	4	1
	4	8	2	1	3	5		6
		5	6			2	9	3
2	6	1	9		7	4	5	
			8		6		2	
				2	5			9
7		6	3		1		8	4
5	1	3	4	9	8	7	6	2
8	9	4	7	6	2			

Puzzle 45
Medium

	9	4	2	8	3	6		5
5		2	6			8	3	4
	6			1				7
	2	5		4	9	1	7	8
4	8	7	1	5	2		6	
1		9	7	6	8	5	4	2
2	5	1	9		4	7	8	6
8			5			2		
	7		8	2		4	5	

Puzzle 46
Medium

7	3	2	9	8	1	6	4	5
8	1	6	5	4	7	2	9	3
4	5	9	3	6	2	8	1	7
6	4	3	2	7	9	1	5	8
5			4	1	6	9	3	2
9	2	1	8	3	5	7		4
2	6							

Puzzle 47
Medium

2	5		3				4	6
		9	4			3	5	
		4	5	6	9	1		
9		5		2	3		7	1
		1	6	4		2	8	
8	2	6	7		1	4	3	5
		7	2	3	8	5	6	4
		3	1	5		7	9	2
5	6	2	9	7		8		

Puzzle 48
Medium

7							2	
4	2					9	5	
6			2				7	3
3	7	9		8	1	5		2
2	8	4		5		7	1	9
5	1	6	9	7	2	3	8	4
8	6					1	9	7
9	3	7	1	6	8	2	4	5
1	4					6	3	8

Puzzle 49
Medium

8	2	5	3	7	4		9	1
		3		2	6	8	4	
	4		8			2	3	
				4			2	8
2			6	8	3	7		9
3		8	2			4		6
5	8			6	2	1	7	3
7	3	2	1	9	8	5	6	4
4	6	1	7	3	5		8	2

Puzzle 50
Medium

	2			1	9	8		
4	9	1	6	5	8	3	7	2
	8	7	4	2	3	1		
			1	8	5		6	
	6		9		7		5	1
			2		6		8	
			5	7	4	9	2	8
2	5	4	8	9	1	6	3	7
	7		3	6	2	5	1	4

Puzzle 51
Hard

	2		4	9		1		5
9	1			3		4	7	2
	6			2			8	9
7		2				8	4	
	3			4	9		2	
1		6		8		5	9	3
	7		9			2		8
	8	9				6		
	5	1	8	7		9	3	4

Puzzle 52
Hard

	1	7	5	6	9	3	4	8
	5		3	4	7	1		
3	4		1	2	8			5
6		5	4	1	3			
4		3	6	9	2		5	1
9	2	1		7	5	4		
5	9	8		3				4
7			9	8				
				5				

Puzzle 53
Hard

7	8	1	6	2	3	4	5	9
4	9	3				6	1	2
6	2	5	9	1	4	8	3	7
		6	2	9				
		2						
	7	9	5				2	
2	6							
9	1			6	2			
					9	2	6	1

Puzzle 54
Hard

8	9	7	1		3	4		6
1	5	6						
2	3	4		6				
9	7	8	2	3	1	5	6	4
4	1	3	5	8	6	2	7	9
5	6	2	7	9	4		3	
	4						9	
	2	9			8			
	8							

Puzzle 55
Hard

4		3		9	5		2	7
	8			2			9	6
7	2		3	8	6			
	7					5	3	9
3	5	8		7				4
9	1				3		7	
	4					9		3
						6	4	5
6	3	5	4	1	9	7	8	2

Puzzle 56
Hard

4	3		6		8	5		
	6				4			3
		1				4		6
7	2	5	4	8	9	3	6	1
9	8	4	1	6	3	7		
6	1	3				9	4	8
3	4	2						
								4
1	7	6	8	4	5	2	3	9

Puzzle 57
Hard

	1			4				
7			6					
		5			2			9
		4			8			2
5		1		2				
8	2	3	4					
1	5	9	2	6	7	4	8	3
2	3	6	8	5	4	7	9	1
4	8	7			9	2	5	6

Puzzle 58
Hard

			4			3	7	2
5	9	3	2	6	7	1	4	8
4	7	2	1	8	3	6		9
			8					6
		6	3			8		
		9	6				3	
			9	4	6	5		
9	1	5	7	3	8	2	6	4
			5	2	1			

Puzzle 59
Hard

	7				5	4	2	
	2		7	3	4		5	
4		8		6				1
	4	7			6	2		5
8		2	3	5			4	
		5	4		7	8		
	8		5	4		7	6	9
				1	8	5	3	
		4		7	9	1	8	2

Puzzle 60
Hard

1	7		4		5			
4					7	1		5
5			1		2	4		
	1	9	5	2	6	3	4	
		4	3		1	8		
	3	7			4			1
9	8	6	2		3	7	1	4
3	4	5	7	1		2		
7	2	1						

Puzzle 61
Hard

		8	9	2		6	3	4
6	9	2	4	8	3	1	7	5
3			5	6		9	8	2
			7	6			9	
				9	4			
9	3			5	8	4		
		9	8		2			
		1	6		9		5	
		3	7		5			9

Puzzle 62
Hard

			8				4	2
	2	5	9	4			8	6
		4	2				5	3
	4	8		2		3		9
1			4		8	2		5
2					7	8		4
					4	6		8
					9	4		7
4	6	3	7	8	2	5	9	1

Puzzle 63
Hard

3			4		8	1	7	
			5	3	1	8		
	1		7			5	3	
			9	4	6	2		3
2	3	4	1	5	7	9	6	8
			2	8	3	4		7
			6	7		3		
			3	1		7	8	
		3	8	2	5	6	4	1

Puzzle 64
Hard

5	1						2	9
7	8		2	9		5	1	4
4	2	9					8	3
3		1				2	9	
6							7	
2		8				3	4	
9	6		5			1	3	7
8	3	7	6	1	9	4	5	
1						9	6	8

Puzzle 65
Hard

				8	1		3	
8	5	7	3		4	1		
3			6		5			7
	7	2	4	5				
		3	8		7			5
6	8	5	2	1	9	4		3
5		4	1	7				
2			9	3		7	5	4
7			5	4				

Puzzle 66
Hard

2		8			6	9	5	
						8		2
			2	8		6		7
			9			3	2	1
7			1	3	8		9	6
1		3					7	8
5	3	7	8			2	6	9
			7	6	3	1	8	5
6	8	1	5	9	2	7		

Puzzle 67
Hard

	8		7		3			
	5	2	9		8			7
	7		5		2			8
	9	5	6	3				
8	6		2	9				5
	2	1	8	5				9
5	3	8	4	2	9		7	
6	4	9	1	7	5			3
2	1	7	3		6			4

Puzzle 68
Hard

3	4	8	2	5			6	9
		5	9	6		8	4	3
1	9		8			2		
				1		7		
7	2	1	3	8	6	4		5
5	8			9	4	6		
	1		4	2		3	7	6
	5	7		3	8		2	1
6	3							

Puzzle 69
Hard

2	9		7	6		5	8	
7	8							
	6							7
3		9	6	1	2	7		8
6		7	8	3	4	9	1	
8		1	9	5	7		3	
	3	2				8		
	1	8	2	4		3	7	
9	7	6		8				

Puzzle 70
Hard

3	1	5	6		8	9	7	4
8	6	4	9	5	7	1		3
2	7	9	4	1	3			
			8					
6	4		7	9	1	2		
9	5		3		6	8	4	
4				8	9			
7	8		1	4				

Puzzle 71
Hard

			4	9	1	5		
		4	5	6	2		9	1
5	9	1	7	3	8	2	4	6
			9	8		4		2
	4		2	7			3	
2	5		1	4	3			9
			6	5			2	
3			8	1				
			3	2			1	

Puzzle 72
Hard

		1	2			8		
					8			
6		3	9	1	4			
					9	4	6	
					3	7	1	8
7		4		8	5	3		
1	3	7			6	2	5	
4	5		3		2	6	8	1
8	2	6	5	4	1	9	7	3

Puzzle 73
Hard

1	9	7	3	8				
3				4				1
6		4	5	1				3
9				6		3		2
	6	3		2	8			9
				3				8
8	3	1		7		9	2	5
2	7	9	8	5	3			
5	4	6	2	9		8	3	7

Puzzle 74
Hard

8	7	4				2	1	5
			1	4	2		7	
2	1		7	8		6	4	9
4	6		2	5	3	9	8	1
3						7	2	4
1	8	2	9	7	4	5	3	6
				2			6	
	2						5	
						1	9	2

Puzzle 75
Hard

	4	9	2		3			
				9		3		
	3	1	7		5		9	8
		2	4	3	1		5	
4	5	7	6	9	8	1	2	3
3	1	6	5	7	2	8	4	9
		3					6	
		9					3	
		4	3	5			8	

Puzzle 76
Fiendish

		2		6				4
		3	9	2				
			1	5	4			
				1				5
2			4	3	5			
			8					
				4	3	1	5	8
3	8	4	5	9	1	6	7	2
		5	8	7		4		

Puzzle 77
Fiendish

4	2	5	6	1	9	7	3	8
6			3	2	4			5
		3	7	8	5			
7		1	5	9		3	2	4
5		2						
		9	2					
				3			6	7
			4	5				

Puzzle 78
Fiendish

4	6		3	8	1			
2	1	8				6	3	4
			4	6	2			
			6		4			
6		2					4	
9	3	4	8		7	5	6	
		6						
		5			6			
	2		5	4	8	9	7	6

Puzzle 79
Fiendish

			3		8			
			5		1			
	1	5						
2	3	7	8	5	9	4	6	
1	9	6						
5		4	6	1				
		1	9		7		8	
7		3	2	8	5			
	5		1					

Puzzle 80
Fiendish

			7			3	5	1
1	3							
5	7		1		3			4
8	5	7	3	4	6	9	1	2
			2	1	8	7	3	5
3	1	2	5			4	6	8
	5							

Puzzle 81
Fiendish

8	3	5	9	4	1	7	2	6
			6	8	3	4		5
	6	4				8	1	3
5	4		1	6	9			8
	8	6			5	1		9
	1	9				6	5	

Puzzle 82
Fiendish

	1	4	3				2	
		6			4		3	
3	9	2						
				3				
	6			4		3		8
		3				2		
9	4	7	8					
2	5	1	9	6	3	7	8	4
6	3	8	4					2

Puzzle 83
Fiendish

7				4				2
4	9	8	6	7	2	1	3	5
2	6			8	1	9	4	7
9		2	1					
	8		2	9				
1						2		
5	2							
	1			5		7	2	
				2			5	1

Puzzle 84
Fiendish

		8		3		1		
		5				8		4
	6			8	2	7		
				6		2		
		2				5		
6								
		6	8	5	9	4		2
		1	7	4	6			
5	4	9	2	1	3	6	8	7

Puzzle 85
Fiendish

2	3	6				1	8	4
9	5	4	6	8	1		2	
8	7		4	3	2	9	6	5
							5	
	2					6		
			2		6	4		
3	2			6		8		
	8	5	1					
					3	5		

Puzzle 86
Fiendish

7		3	6	9	2		4	8
		9	4					
				1			2	9
		5				8	6	
			5			3	1	
3						9	5	7
9	4						7	
6	3		9			4	8	
5	7					2	9	3

Puzzle 87
Fiendish

1	9	8	4					
	5		3			8	9	
		7	9					
		6			9		5	
	1		7			2		
			6		3			1
2	4	3	8	9				
		1	5	3	4			
	8	9					3	4

Puzzle 88
Fiendish

3							9	6
		9						7
			9			3		4
		2		9	5	7	3	8
					7		6	1
			3	6			4	
		4	2	5		6		9
2	8	5	6	1	9	4	7	3

Puzzle 89
Fiendish

7	4		8	6				
8		1						
2				7				8
1	5		6	3	7	4	8	
			1	5	8	7		
3	7	8	2	4	9	6	1	5
6								
			3				2	
	1							

Puzzle 90
Fiendish

3				1				
2		7		4			9	
6			2	5				
		1				7		
4	7							
	2			7			8	
7			1	8	4			2
8			7	2	5	4		
1	4	2	9	3	6	5	7	8

Puzzle 91
Fiendish

	3	8				9		
4	9	1	5	8	3	2	7	6
				9	1	8		
		4				1		
			8					5
	1		3	4	8	5		7
8	4	9	1	5	7	3	6	
	5	7				4		

Puzzle 92
Fiendish

9	5	2	8			1	4	6
6					9		2	7
								5
2	7		4		5	8		
1	6	5			7	4		
8	4							
5	8	4						
7	2						5	4
3	9	6	7	5				

Puzzle 93
Fiendish

1								
		5		1				
4		2	6	9	8		7	
3	1	6	2	7	5			
8	7	9	1	4	3	6	5	2
5	8	7	4	6			2	
6	3	1	7			4		
		4						

Puzzle 94
Fiendish

		2	1		7	3		
5		8					6	1
		3		6				4
		9						
		5					7	
8		1						
	5	6				4	3	
3	8	4	5		6			7
2		7			4	8	5	6

Puzzle 95
Fiendish

	4	7	2	8	1			
3			9	6	5	8		4
		8		4	7			
			6	9	8		5	1
		6				7	9	3
4	9	5	1	7	3	6		
2			8	3				
		9						
			7	1				

Puzzle 96
Fiendish

						9		
8		9			1	4		
				3	9			
6			2				9	
		5		9				
9		4			8			
4	9	1	5	8	2	7	3	6
	7	8	3	4	6	1		9
3			9	1	7			

Puzzle 97
Fiendish

					7		1	6
					3	9	8	5
					5			
				3	6			8
			9		8			1
	8			4				9
6		8					9	
2	4	3	6	8	9	1	5	7
1			3		4	8	6	2

Puzzle 98
Fiendish

2				4	3	1		
							9	
7	3		1				5	
1	2	5	6	3			7	9
	4				7		3	1
	1	3	4	7			2	
6	8	2	9	1	5	7	4	3
			3				1	

Puzzle 99
Fiendish

2							1	5
							3	6
5					7		9	8
			2	5		1	7	9
				3	1	6	4	2
	2	1				8	5	3
	5	6	4		9	3		1
3						5		4
						9		7

Puzzle 100
Fiendish

				7		4	3	
	3				9		8	
5		8		3		7	9	2
8	2	6	7	9		5	1	3
4	1	9	3				2	
3	7	5					6	4
2								
				8		2	7	

Solutions

Puzzle 1

7	8	3	1	2	9	4	5	6
1	6	9	7	5	4	3	8	2
2	4	5	8	3	6	9	1	7
5	1	4	9	6	8	7	2	3
6	2	8	3	7	1	5	4	9
9	3	7	2	4	5	1	6	8
4	7	2	6	1	3	8	9	5
8	5	6	4	9	7	2	3	1
3	9	1	5	8	2	6	7	4

Puzzle 2

1	4	6	3	2	8	9	5	7
5	9	2	6	7	4	8	1	3
3	8	7	9	1	5	2	6	4
4	5	1	2	6	9	3	7	8
2	3	8	1	5	7	4	9	6
7	6	9	8	4	3	1	2	5
6	2	5	4	3	1	7	8	9
9	7	3	5	8	2	6	4	1
8	1	4	7	9	6	5	3	2

Puzzle 3

8	6	5	1	7	4	3	2	9
9	7	1	3	6	2	4	8	5
2	4	3	5	9	8	1	6	7
4	5	7	8	2	3	6	9	1
3	8	2	9	1	6	5	7	4
6	1	9	7	4	5	2	3	8
7	9	4	2	3	1	8	5	6
1	3	8	6	5	9	7	4	2
5	2	6	4	8	7	9	1	3

Puzzle 4

1	5	8	6	2	4	7	9	3
9	7	2	1	3	8	4	5	6
6	4	3	7	5	9	2	1	8
8	6	9	2	4	7	1	3	5
4	2	1	5	6	3	9	8	7
5	3	7	8	9	1	6	4	2
2	1	6	4	8	5	3	7	9
3	8	4	9	7	2	5	6	1
7	9	5	3	1	6	8	2	4

Puzzle 5

4	8	6	1	5	2	9	7	3
9	5	7	4	3	8	1	6	2
2	1	3	9	6	7	5	4	8
1	9	2	8	7	4	3	5	6
5	6	8	3	2	9	7	1	4
3	7	4	6	1	5	8	2	9
6	2	1	5	8	3	4	9	7
8	4	5	7	9	6	2	3	1
7	3	9	2	4	1	6	8	5

Puzzle 6

1	3	4	5	8	7	9	2	6
2	6	5	9	4	3	7	1	8
7	9	8	2	6	1	3	4	5
9	2	7	4	5	6	8	3	1
4	1	3	8	7	9	5	6	2
5	8	6	1	3	2	4	9	7
8	7	2	6	9	4	1	5	3
3	4	1	7	2	5	6	8	9
6	5	9	3	1	8	2	7	4

Puzzle 7

3	6	8	1	2	4	7	5	9
9	7	5	6	8	3	2	1	4
2	1	4	9	5	7	8	6	3
8	3	1	4	6	2	9	7	5
6	5	9	3	7	8	4	2	1
7	4	2	5	1	9	3	8	6
1	9	7	2	4	6	5	3	8
4	8	6	7	3	5	1	9	2
5	2	3	8	9	1	6	4	7

Puzzle 8

5	1	2	6	4	8	3	7	9
8	3	4	9	7	1	6	5	2
9	6	7	3	2	5	8	1	4
7	8	3	2	5	6	4	9	1
6	4	5	1	8	9	7	2	3
2	9	1	7	3	4	5	6	8
1	5	8	4	9	7	2	3	6
4	2	9	5	6	3	1	8	7
3	7	6	8	1	2	9	4	5

Puzzle 9

9	3	7	6	1	2	4	5	8
6	8	4	5	9	7	3	2	1
5	2	1	4	3	8	6	7	9
1	4	3	8	7	9	2	6	5
7	9	8	2	5	6	1	4	3
2	5	6	1	4	3	9	8	7
3	6	9	7	2	5	8	1	4
8	1	5	9	6	4	7	3	2
4	7	2	3	8	1	5	9	6

Puzzle 10

6	2	9	1	8	7	3	4	5
3	5	7	4	2	6	8	1	9
4	8	1	9	5	3	7	6	2
2	9	4	7	6	1	5	8	3
5	7	6	2	3	8	1	9	4
8	1	3	5	9	4	2	7	6
9	6	2	8	7	5	4	3	1
7	4	5	3	1	9	6	2	8
1	3	8	6	4	2	9	5	7

Puzzle 11

4	3	5	8	2	9	1	6	7
8	6	2	3	1	7	5	9	4
1	7	9	6	4	5	8	3	2
7	5	8	2	3	1	6	4	9
9	4	3	5	7	6	2	8	1
2	1	6	9	8	4	3	7	5
6	2	7	4	5	8	9	1	3
5	9	4	1	6	3	7	2	8
3	8	1	7	9	2	4	5	6

Puzzle 12

6	9	3	8	1	4	2	7	5
5	2	8	6	7	9	3	1	4
7	4	1	5	2	3	6	9	8
2	5	6	1	4	8	7	3	9
4	3	7	9	5	2	1	8	6
8	1	9	3	6	7	4	5	2
3	7	5	2	8	6	9	4	1
9	8	2	4	3	1	5	6	7
1	6	4	7	9	5	8	2	3

Puzzle 13

7	1	8	9	4	2	5	6	3
4	2	6	3	1	5	9	7	8
3	5	9	8	7	6	1	4	2
1	8	7	2	9	4	3	5	6
5	3	2	6	8	7	4	9	1
9	6	4	5	3	1	2	8	7
2	4	5	7	6	3	8	1	9
6	9	3	1	5	8	7	2	4
8	7	1	4	2	9	6	3	5

Puzzle 14

1	3	7	6	8	4	2	5	9
8	2	6	1	9	5	4	7	3
9	4	5	3	2	7	6	8	1
4	5	9	2	3	6	7	1	8
7	1	8	5	4	9	3	6	2
2	6	3	7	1	8	9	4	5
6	8	1	9	7	3	5	2	4
3	7	2	4	5	1	8	9	6
5	9	4	8	6	2	1	3	7

Puzzle 15

3	6	7	8	1	5	4	2	9
5	8	2	4	7	9	1	6	3
1	9	4	3	2	6	7	8	5
6	5	3	1	9	7	2	4	8
7	2	8	6	5	4	9	3	1
4	1	9	2	8	3	6	5	7
2	7	5	9	6	8	3	1	4
9	3	6	5	4	1	8	7	2
8	4	1	7	3	2	5	9	6

Puzzle 16

3	1	9	8	5	4	6	7	2
5	4	6	1	2	7	3	8	9
7	2	8	6	3	9	4	5	1
9	7	4	5	8	3	2	1	6
8	5	2	9	1	6	7	4	3
1	6	3	4	7	2	5	9	8
4	8	7	2	6	1	9	3	5
2	9	5	3	4	8	1	6	7
6	3	1	7	9	5	8	2	4

Puzzle 17

2	1	8	9	5	6	4	3	7
3	7	9	2	4	1	6	8	5
5	6	4	7	3	8	2	1	9
8	9	2	1	6	7	3	5	4
6	5	1	3	8	4	7	9	2
7	4	3	5	2	9	1	6	8
4	2	6	8	9	3	5	7	1
9	3	7	4	1	5	8	2	6
1	8	5	6	7	2	9	4	3

Puzzle 18

6	2	1	4	5	8	7	9	3
9	7	4	6	1	3	5	2	8
8	5	3	2	7	9	4	1	6
3	8	7	1	9	5	2	6	4
2	1	6	7	8	4	9	3	5
4	9	5	3	2	6	1	8	7
7	3	9	8	4	1	6	5	2
1	4	8	5	6	2	3	7	9
5	6	2	9	3	7	8	4	1

Puzzle 19

4	6	1	3	5	8	7	2	9
9	8	5	1	7	2	6	3	4
7	2	3	4	9	6	5	1	8
5	7	8	6	2	9	3	4	1
6	1	4	5	8	3	2	9	7
3	9	2	7	4	1	8	5	6
1	4	6	8	3	5	9	7	2
8	3	9	2	1	7	4	6	5
2	5	7	9	6	4	1	8	3

Puzzle 20

6	2	3	1	4	8	7	9	5
8	4	7	9	3	5	1	6	2
1	5	9	7	6	2	4	8	3
2	7	1	4	5	6	8	3	9
4	6	8	3	7	9	2	5	1
3	9	5	2	8	1	6	7	4
5	8	2	6	9	4	3	1	7
9	3	4	8	1	7	5	2	6
7	1	6	5	2	3	9	4	8

Puzzle 21

6	1	2	7	4	8	3	9	5
9	7	4	2	5	3	8	6	1
8	5	3	1	6	9	7	2	4
1	8	5	6	9	2	4	3	7
2	3	9	8	7	4	5	1	6
4	6	7	3	1	5	2	8	9
7	2	6	4	3	1	9	5	8
3	9	1	5	8	7	6	4	2
5	4	8	9	2	6	1	7	3

Puzzle 22

4	5	2	6	1	8	3	7	9
9	3	6	7	2	4	5	8	1
7	8	1	9	3	5	2	6	4
2	4	5	3	6	9	8	1	7
6	7	8	2	4	1	9	5	3
1	9	3	8	5	7	6	4	2
8	2	7	4	9	6	1	3	5
5	6	9	1	7	3	4	2	8
3	1	4	5	8	2	7	9	6

Puzzle 23

4	6	7	1	9	5	2	8	3
1	3	8	7	4	2	9	6	5
9	5	2	6	3	8	1	7	4
5	2	9	4	8	6	7	3	1
7	4	6	3	1	9	8	5	2
8	1	3	5	2	7	4	9	6
2	9	1	8	6	3	5	4	7
6	7	4	9	5	1	3	2	8
3	8	5	2	7	4	6	1	9

Puzzle 24

7	1	4	6	9	2	3	5	8
3	8	9	7	5	4	6	2	1
6	5	2	1	3	8	7	9	4
2	7	8	3	6	1	9	4	5
4	6	5	9	8	7	2	1	3
9	3	1	2	4	5	8	6	7
1	4	7	8	2	9	5	3	6
5	2	6	4	7	3	1	8	9
8	9	3	5	1	6	4	7	2

Puzzle 25

1	8	7	4	3	9	6	5	2
2	5	4	7	6	1	3	9	8
9	6	3	5	2	8	7	1	4
5	4	8	9	7	6	1	2	3
6	3	2	1	4	5	9	8	7
7	1	9	3	8	2	5	4	6
4	9	6	2	1	7	8	3	5
3	7	5	8	9	4	2	6	1
8	2	1	6	5	3	4	7	9

Puzzle 26

4	8	7	2	1	5	9	3	6
1	6	3	7	9	4	5	8	2
2	5	9	3	6	8	7	4	1
7	4	1	6	3	2	8	5	9
5	9	6	4	8	7	1	2	3
3	2	8	1	5	9	6	7	4
8	7	2	9	4	6	3	1	5
6	3	5	8	2	1	4	9	7
9	1	4	5	7	3	2	6	8

Puzzle 27

5	1	7	3	4	8	6	9	2
2	6	4	9	5	7	1	8	3
8	9	3	2	1	6	5	4	7
3	8	9	5	6	1	7	2	4
7	5	1	4	3	2	8	6	9
4	2	6	8	7	9	3	1	5
6	3	2	1	9	5	4	7	8
9	7	5	6	8	4	2	3	1
1	4	8	7	2	3	9	5	6

Puzzle 28

4	2	3	1	5	8	9	6	7
5	9	1	3	6	7	2	8	4
7	6	8	9	4	2	5	1	3
3	8	9	6	7	1	4	2	5
1	7	2	5	9	4	8	3	6
6	4	5	2	8	3	7	9	1
8	5	6	7	1	9	3	4	2
2	1	4	8	3	5	6	7	9
9	3	7	4	2	6	1	5	8

Puzzle 29

7	2	5	1	3	8	4	9	6
9	4	1	2	6	5	8	7	3
6	8	3	4	9	7	2	5	1
2	9	7	8	1	6	5	3	4
3	1	8	5	4	9	6	2	7
5	6	4	3	7	2	1	8	9
4	5	9	6	8	3	7	1	2
1	3	2	7	5	4	9	6	8
8	7	6	9	2	1	3	4	5

Puzzle 30

7	3	6	4	5	2	9	8	1
9	2	1	6	7	8	5	3	4
5	4	8	1	9	3	7	2	6
3	8	4	7	2	1	6	9	5
2	6	9	8	4	5	1	7	3
1	5	7	9	3	6	8	4	2
4	7	5	3	1	9	2	6	8
8	9	2	5	6	4	3	1	7
6	1	3	2	8	7	4	5	9

Puzzle 31

2	8	9	4	7	1	5	6	3
7	5	4	9	3	6	1	2	8
6	3	1	2	5	8	9	7	4
9	2	3	8	1	7	4	5	6
1	7	6	5	4	2	3	8	9
5	4	8	3	6	9	7	1	2
4	9	7	6	8	5	2	3	1
3	6	5	1	2	4	8	9	7
8	1	2	7	9	3	6	4	5

Puzzle 32

3	4	7	6	9	8	2	1	5
8	5	6	7	2	1	4	9	3
2	1	9	3	5	4	8	7	6
1	9	3	8	7	5	6	2	4
7	6	4	1	3	2	9	5	8
5	8	2	4	6	9	7	3	1
4	7	1	2	8	3	5	6	9
6	3	5	9	4	7	1	8	2
9	2	8	5	1	6	3	4	7

Puzzle 33

3	2	5	4	9	1	7	8	6
7	4	8	5	6	2	3	9	1
6	1	9	8	7	3	4	5	2
4	6	2	1	5	9	8	7	3
9	5	3	6	8	7	2	1	4
8	7	1	3	2	4	9	6	5
5	3	6	9	4	8	1	2	7
1	9	7	2	3	6	5	4	8
2	8	4	7	1	5	6	3	9

Puzzle 34

5	7	4	8	1	3	9	2	6
3	8	1	6	9	2	7	4	5
9	2	6	4	5	7	3	8	1
6	9	8	2	4	1	5	7	3
2	5	7	3	6	8	1	9	4
4	1	3	5	7	9	8	6	2
1	6	2	9	8	5	4	3	7
8	4	5	7	3	6	2	1	9
7	3	9	1	2	4	6	5	8

Puzzle 35

2	3	6	1	9	7	5	8	4
4	7	9	3	5	8	1	2	6
8	1	5	4	2	6	3	9	7
5	8	1	2	4	3	7	6	9
9	2	7	8	6	5	4	3	1
3	6	4	9	7	1	2	5	8
6	4	2	5	1	9	8	7	3
7	5	3	6	8	4	9	1	2
1	9	8	7	3	2	6	4	5

Puzzle 36

4	5	9	1	6	8	3	7	2
8	1	3	2	7	4	9	6	5
6	7	2	5	9	3	1	8	4
2	8	1	4	3	7	5	9	6
7	4	6	9	5	2	8	1	3
9	3	5	6	8	1	2	4	7
3	9	4	7	1	5	6	2	8
5	6	7	8	2	9	4	3	1
1	2	8	3	4	6	7	5	9

Puzzle 37

1	8	7	5	9	4	2	3	6
2	9	6	8	3	7	1	4	5
5	4	3	1	6	2	9	8	7
3	7	2	9	1	5	8	6	4
8	1	9	4	7	6	3	5	2
4	6	5	3	2	8	7	9	1
9	5	4	2	8	1	6	7	3
7	2	8	6	5	3	4	1	9
6	3	1	7	4	9	5	2	8

Puzzle 38

6	4	3	1	5	8	7	9	2
9	8	7	4	6	2	3	5	1
5	2	1	7	9	3	6	4	8
1	3	9	5	2	4	8	6	7
4	7	2	8	1	6	5	3	9
8	5	6	3	7	9	1	2	4
2	9	5	6	8	7	4	1	3
3	6	8	2	4	1	9	7	5
7	1	4	9	3	5	2	8	6

Puzzle 39

9	3	1	7	4	5	2	6	8
8	4	2	9	1	6	3	5	7
7	5	6	2	3	8	9	1	4
1	9	4	5	8	2	7	3	6
2	6	3	4	9	7	1	8	5
5	8	7	3	6	1	4	2	9
6	2	9	8	7	3	5	4	1
3	7	8	1	5	4	6	9	2
4	1	5	6	2	9	8	7	3

Puzzle 40

5	1	7	4	8	9	6	2	3
9	2	4	6	3	5	1	7	8
6	3	8	1	7	2	4	9	5
7	6	2	9	4	3	8	5	1
8	9	5	7	1	6	3	4	2
3	4	1	5	2	8	7	6	9
1	7	9	3	5	4	2	8	6
4	8	6	2	9	1	5	3	7
2	5	3	8	6	7	9	1	4

Puzzle 41

5	3	8	9	7	2	6	1	4
6	1	2	4	3	8	7	5	9
9	4	7	5	6	1	2	8	3
2	5	1	7	4	9	3	6	8
7	9	3	8	2	6	1	4	5
8	6	4	1	5	3	9	7	2
1	8	5	2	9	7	4	3	6
3	7	9	6	8	4	5	2	1
4	2	6	3	1	5	8	9	7

Puzzle 42

4	5	1	6	9	2	3	8	7
6	8	2	1	3	7	9	4	5
7	3	9	8	5	4	6	1	2
3	9	4	2	7	5	1	6	8
5	1	7	3	8	6	4	2	9
2	6	8	4	1	9	5	7	3
1	7	3	5	6	8	2	9	4
9	4	5	7	2	1	8	3	6
8	2	6	9	4	3	7	5	1

Puzzle 43

8	9	2	1	4	3	7	5	6
7	3	6	9	5	8	4	2	1
1	5	4	6	7	2	3	8	9
5	4	8	2	3	9	6	1	7
3	1	7	5	6	4	2	9	8
6	2	9	8	1	7	5	3	4
4	7	1	3	8	5	9	6	2
2	8	3	4	9	6	1	7	5
9	6	5	7	2	1	8	4	3

Puzzle 44

6	3	2	5	7	9	8	4	1
9	4	8	2	1	3	5	7	6
1	7	5	6	8	4	2	9	3
2	6	1	9	3	7	4	5	8
3	5	9	8	4	6	1	2	7
4	8	7	1	2	5	6	3	9
7	2	6	3	5	1	9	8	4
5	1	3	4	9	8	7	6	2
8	9	4	7	6	2	3	1	5

Puzzle 45

7	9	4	2	8	3	6	1	5
5	1	2	6	9	7	8	3	4
3	6	8	4	1	5	9	2	7
6	2	5	3	4	9	1	7	8
4	8	7	1	5	2	3	6	9
1	3	9	7	6	8	5	4	2
2	5	1	9	3	4	7	8	6
8	4	3	5	7	6	2	9	1
9	7	6	8	2	1	4	5	3

Puzzle 46

7	3	2	9	8	1	6	4	5
8	1	6	5	4	7	2	9	3
4	5	9	3	6	2	8	1	7
6	4	3	2	7	9	1	5	8
5	7	8	4	1	6	9	3	2
9	2	1	8	3	5	7	6	4
2	6	4	1	5	8	3	7	9
1	8	5	7	9	3	4	2	6
3	9	7	6	2	4	5	8	1

Puzzle 47

2	5	8	3	1	7	9	4	6
6	1	9	4	8	2	3	5	7
3	7	4	5	6	9	1	2	8
9	4	5	8	2	3	6	7	1
7	3	1	6	4	5	2	8	9
8	2	6	7	9	1	4	3	5
1	9	7	2	3	8	5	6	4
4	8	3	1	5	6	7	9	2
5	6	2	9	7	4	8	1	3

Puzzle 48

7	5	3	8	1	9	4	2	6
4	2	8	7	3	6	9	5	1
6	9	1	2	4	5	8	7	3
3	7	9	4	8	1	5	6	2
2	8	4	6	5	3	7	1	9
5	1	6	9	7	2	3	8	4
8	6	5	3	2	4	1	9	7
9	3	7	1	6	8	2	4	5
1	4	2	5	9	7	6	3	8

Puzzle 49

8	2	5	3	7	4	6	9	1
1	7	3	9	2	6	8	4	5
9	4	6	8	5	1	2	3	7
6	1	7	5	4	9	3	2	8
2	5	4	6	8	3	7	1	9
3	9	8	2	1	7	4	5	6
5	8	9	4	6	2	1	7	3
7	3	2	1	9	8	5	6	4
4	6	1	7	3	5	9	8	2

Puzzle 50

3	2	6	7	1	9	8	4	5
4	9	1	6	5	8	3	7	2
5	8	7	4	2	3	1	9	6
7	4	9	1	8	5	2	6	3
8	6	2	9	3	7	4	5	1
1	3	5	2	4	6	7	8	9
6	1	3	5	7	4	9	2	8
2	5	4	8	9	1	6	3	7
9	7	8	3	6	2	5	1	4

Puzzle 51

3	2	7	4	9	8	1	6	5
9	1	8	5	3	6	4	7	2
5	6	4	7	2	1	3	8	9
7	9	2	1	5	3	8	4	6
8	3	5	6	4	9	7	2	1
1	4	6	2	8	7	5	9	3
4	7	3	9	6	5	2	1	8
2	8	9	3	1	4	6	5	7
6	5	1	8	7	2	9	3	4

Puzzle 52

2	1	7	5	6	9	3	4	8
8	5	6	3	4	7	1	2	9
3	4	9	1	2	8	6	7	5
6	7	5	4	1	3	8	9	2
4	8	3	6	9	2	7	5	1
9	2	1	8	7	5	4	3	6
5	9	8	7	3	1	2	6	4
7	6	2	9	8	4	5	1	3
1	3	4	2	5	6	9	8	7

Puzzle 53

7	8	1	6	2	3	4	5	9
4	9	3	7	5	8	6	1	2
6	2	5	9	1	4	8	3	7
8	4	6	2	9	1	3	7	5
1	5	2	3	8	7	9	4	6
3	7	9	5	4	6	1	2	8
2	6	8	1	3	5	7	9	4
9	1	7	4	6	2	5	8	3
5	3	4	8	7	9	2	6	1

Puzzle 54

8	9	7	1	2	3	4	5	6
1	5	6	8	4	7	9	2	3
2	3	4	9	6	5	1	8	7
9	7	8	2	3	1	5	6	4
4	1	3	5	8	6	2	7	9
5	6	2	7	9	4	8	3	1
6	4	1	3	5	2	7	9	8
3	2	9	4	7	8	6	1	5
7	8	5	6	1	9	3	4	2

Puzzle 55

4	6	3	1	9	5	8	2	7
5	8	1	7	2	4	3	9	6
7	2	9	3	8	6	4	5	1
2	7	4	8	6	1	5	3	9
3	5	8	9	7	2	1	6	4
9	1	6	5	4	3	2	7	8
8	4	2	6	5	7	9	1	3
1	9	7	2	3	8	6	4	5
6	3	5	4	1	9	7	8	2

Puzzle 56

4	3	9	6	1	8	5	2	7
2	6	7	5	9	4	1	8	3
8	5	1	3	2	7	4	9	6
7	2	5	4	8	9	3	6	1
9	8	4	1	6	3	7	5	2
6	1	3	7	5	2	9	4	8
3	4	2	9	7	6	8	1	5
5	9	8	2	3	1	6	7	4
1	7	6	8	4	5	2	3	9

Puzzle 57

6	1	2	9	4	3	8	7	5
7	9	8	6	1	5	3	2	4
3	4	5	7	8	2	6	1	9
9	6	4	5	7	8	1	3	2
5	7	1	3	2	6	9	4	8
8	2	3	4	9	1	5	6	7
1	5	9	2	6	7	4	8	3
2	3	6	8	5	4	7	9	1
4	8	7	1	3	9	2	5	6

Puzzle 58

1	6	8	4	5	9	3	7	2
5	9	3	2	6	7	1	4	8
4	7	2	1	8	3	6	5	9
3	2	1	8	7	5	4	9	6
7	5	6	3	9	4	8	2	1
8	4	9	6	1	2	7	3	5
2	8	7	9	4	6	5	1	3
9	1	5	7	3	8	2	6	4
6	3	4	5	2	1	9	8	7

Puzzle 59

9	7	3	1	8	5	4	2	6
1	2	6	7	3	4	9	5	8
4	5	8	9	6	2	3	7	1
3	4	7	8	9	6	2	1	5
8	9	2	3	5	1	6	4	7
6	1	5	4	2	7	8	9	3
2	8	1	5	4	3	7	6	9
7	6	9	2	1	8	5	3	4
5	3	4	6	7	9	1	8	2

Puzzle 60

1	7	8	4	3	5	6	2	9
4	6	2	8	9	7	1	3	5
5	9	3	1	6	2	4	7	8
8	1	9	5	2	6	3	4	7
6	5	4	3	7	1	8	9	2
2	3	7	9	8	4	5	6	1
9	8	6	2	5	3	7	1	4
3	4	5	7	1	9	2	8	6
7	2	1	6	4	8	9	5	3

Puzzle 61

5	7	8	9	2	1	6	3	4
6	9	2	4	8	3	1	7	5
3	1	4	5	6	7	9	8	2
1	4	5	2	7	6	3	9	8
8	2	6	3	9	4	5	1	7
9	3	7	1	5	8	4	2	6
4	5	9	8	3	2	7	6	1
7	8	1	6	4	9	2	5	3
2	6	3	7	1	5	8	4	9

Puzzle 62

9	7	6	8	5	3	1	4	2
3	2	5	9	4	1	7	8	6
8	1	4	2	7	6	9	5	3
6	4	8	1	2	5	3	7	9
1	3	7	4	9	8	2	6	5
2	5	9	3	6	7	8	1	4
7	9	1	5	3	4	6	2	8
5	8	2	6	1	9	4	3	7
4	6	3	7	8	2	5	9	1

Puzzle 63

3	5	9	4	6	8	1	7	2
6	2	7	5	3	1	8	9	4
4	1	8	7	9	2	5	3	6
7	8	5	9	4	6	2	1	3
2	3	4	1	5	7	9	6	8
1	9	6	2	8	3	4	5	7
8	4	1	6	7	9	3	2	5
5	6	2	3	1	4	7	8	9
9	7	3	8	2	5	6	4	1

Puzzle 64

5	1	3	4	8	7	6	2	9
7	8	6	2	9	3	5	1	4
4	2	9	1	5	6	7	8	3
3	7	1	8	6	4	2	9	5
6	9	4	3	2	5	8	7	1
2	5	8	9	7	1	3	4	6
9	6	2	5	4	8	1	3	7
8	3	7	6	1	9	4	5	2
1	4	5	7	3	2	9	6	8

Puzzle 65

4	2	6	7	8	1	5	3	9
8	5	7	3	9	4	1	2	6
3	1	9	6	2	5	8	4	7
1	7	2	4	5	3	6	9	8
9	4	3	8	6	7	2	1	5
6	8	5	2	1	9	4	7	3
5	3	4	1	7	6	9	8	2
2	6	1	9	3	8	7	5	4
7	9	8	5	4	2	3	6	1

Puzzle 66

2	1	8	3	7	6	9	5	4
3	7	6	4	5	9	8	1	2
4	5	9	2	8	1	6	3	7
8	6	5	9	4	7	3	2	1
7	2	4	1	3	8	5	9	6
1	9	3	6	2	5	4	7	8
5	3	7	8	1	4	2	6	9
9	4	2	7	6	3	1	8	5
6	8	1	5	9	2	7	4	3

Puzzle 67

1	8	4	7	6	3	5	9	2
3	5	2	9	1	8	4	6	7
9	7	6	5	4	2	3	1	8
4	9	5	6	3	7	2	8	1
8	6	3	2	9	1	7	4	5
7	2	1	8	5	4	6	3	9
5	3	8	4	2	9	1	7	6
6	4	9	1	7	5	8	2	3
2	1	7	3	8	6	9	5	4

Puzzle 68

3	4	8	2	5	7	1	6	9
2	7	5	9	6	1	8	4	3
1	9	6	8	4	3	2	5	7
9	6	4	5	1	2	7	3	8
7	2	1	3	8	6	4	9	5
5	8	3	7	9	4	6	1	2
8	1	9	4	2	5	3	7	6
4	5	7	6	3	8	9	2	1
6	3	2	1	7	9	5	8	4

Puzzle 69

2	9	4	7	6	1	5	8	3
7	8	3	4	2	5	1	6	9
1	6	5	3	9	8	2	4	7
3	4	9	6	1	2	7	5	8
6	5	7	8	3	4	9	1	2
8	2	1	9	5	7	6	3	4
4	3	2	1	7	6	8	9	5
5	1	8	2	4	9	3	7	6
9	7	6	5	8	3	4	2	1

Puzzle 70

3	1	5	6	2	8	9	7	4
8	6	4	9	5	7	1	2	3
2	7	9	4	1	3	5	6	8
1	2	7	8	3	5	4	9	6
6	4	3	7	9	1	2	8	5
5	9	8	2	6	4	7	3	1
9	5	1	3	7	6	8	4	2
4	3	2	5	8	9	6	1	7
7	8	6	1	4	2	3	5	9

Puzzle 71

6	2	8	4	9	1	5	7	3
7	3	4	5	6	2	8	9	1
5	9	1	7	3	8	2	4	6
1	7	3	9	8	5	4	6	2
8	4	9	2	7	6	1	3	5
2	5	6	1	4	3	7	8	9
4	1	7	6	5	9	3	2	8
3	6	2	8	1	4	9	5	7
9	8	5	3	2	7	6	1	4

Puzzle 72

9	4	1	2	5	7	8	3	6
2	7	5	6	3	8	1	4	9
6	8	3	9	1	4	5	2	7
3	1	8	7	2	9	4	6	5
5	9	2	4	6	3	7	1	8
7	6	4	1	8	5	3	9	2
1	3	7	8	9	6	2	5	4
4	5	9	3	7	2	6	8	1
8	2	6	5	4	1	9	7	3

Puzzle 73

1	9	7	3	8	2	5	6	4
3	5	8	6	4	7	2	9	1
6	2	4	5	1	9	7	8	3
9	8	5	1	6	4	3	7	2
4	6	3	7	2	8	1	5	9
7	1	2	9	3	5	6	4	8
8	3	1	4	7	6	9	2	5
2	7	9	8	5	3	4	1	6
5	4	6	2	9	1	8	3	7

Puzzle 74

8	7	4	3	6	9	2	1	5
6	5	9	1	4	2	3	7	8
2	1	3	7	8	5	6	4	9
4	6	7	2	5	3	9	8	1
3	9	5	6	1	8	7	2	4
1	8	2	9	7	4	5	3	6
9	3	8	5	2	1	4	6	7
7	2	1	4	9	6	8	5	3
5	4	6	8	3	7	1	9	2

Puzzle 75

7	4	9	2	8	3	5	1	6
6	8	5	1	4	9	3	7	2
2	3	1	7	6	5	4	9	8
8	9	2	4	3	1	6	5	7
4	5	7	6	9	8	1	2	3
3	1	6	5	7	2	8	4	9
1	7	3	8	2	4	9	6	5
5	2	8	9	1	6	7	3	4
9	6	4	3	5	7	2	8	1

Puzzle 76

5	9	2	3	6	7	8	1	4
1	4	3	9	2	8	5	6	7
7	6	8	1	5	4	2	3	9
8	3	6	2	1	9	7	4	5
2	7	1	4	3	5	9	8	6
4	5	9	7	8	6	3	2	1
9	2	7	6	4	3	1	5	8
3	8	4	5	9	1	6	7	2
6	1	5	8	7	2	4	9	3

Puzzle 77

4	2	5	6	1	9	7	3	8
6	8	7	3	2	4	1	9	5
1	9	3	7	8	5	2	4	6
7	6	1	5	9	8	3	2	4
5	3	2	1	4	6	8	7	9
8	4	9	2	7	3	6	5	1
2	5	8	9	6	7	4	1	3
9	1	4	8	3	2	5	6	7
3	7	6	4	5	1	9	8	2

Puzzle 78

4	6	7	3	8	1	2	5	9
2	1	8	7	5	9	6	3	4
3	5	9	4	6	2	7	1	8
5	7	1	6	2	4	8	9	3
6	8	2	9	3	5	1	4	7
9	3	4	8	1	7	5	6	2
8	9	6	1	7	3	4	2	5
7	4	5	2	9	6	3	8	1
1	2	3	5	4	8	9	7	6

Puzzle 79

6	4	2	3	7	8	1	5	9
3	7	8	5	9	1	2	4	6
9	1	5	4	6	2	8	7	3
2	3	7	8	5	9	4	6	1
1	9	6	7	2	4	5	3	8
5	8	4	6	1	3	7	9	2
4	2	1	9	3	7	6	8	5
7	6	3	2	8	5	9	1	4
8	5	9	1	4	6	3	2	7

Puzzle 80

2	9	8	7	6	4	3	5	1
1	3	4	8	5	2	6	7	9
5	7	6	1	9	3	2	8	4
8	5	7	3	4	6	9	1	2
4	6	9	2	1	8	7	3	5
3	1	2	5	7	9	4	6	8
9	2	5	6	3	1	8	4	7
6	4	1	9	8	7	5	2	3
7	8	3	4	2	5	1	9	6

Puzzle 81

8	3	5	9	4	1	7	2	6
1	2	7	6	8	3	4	9	5
9	6	4	5	7	2	8	1	3
5	4	2	1	6	9	3	7	8
3	8	6	7	2	5	1	4	9
7	1	9	4	3	8	6	5	2
2	7	1	3	5	6	9	8	4
4	5	3	8	9	7	2	6	1
6	9	8	2	1	4	5	3	7

Puzzle 82

5	1	4	3	7	6	8	2	9
8	7	6	2	9	4	1	3	5
3	9	2	5	8	1	4	6	7
4	2	5	1	3	8	9	7	6
1	6	9	7	4	2	3	5	8
7	8	3	6	5	9	2	4	1
9	4	7	8	2	5	6	1	3
2	5	1	9	6	3	7	8	4
6	3	8	4	1	7	5	9	2

Puzzle 83

7	3	1	5	4	9	6	8	2
4	9	8	6	7	2	1	3	5
2	6	5	3	8	1	9	4	7
9	5	2	1	3	7	8	6	4
6	8	7	2	9	4	5	1	3
1	4	3	8	6	5	2	7	9
5	2	4	7	1	6	3	9	8
3	1	9	4	5	8	7	2	6
8	7	6	9	2	3	4	5	1

Puzzle 84

7	9	8	5	3	4	1	2	6
2	3	5	6	7	1	8	9	4
1	6	4	9	8	2	7	3	5
9	1	3	4	6	5	2	7	8
4	8	2	1	9	7	5	6	3
6	5	7	3	2	8	9	4	1
3	7	6	8	5	9	4	1	2
8	2	1	7	4	6	3	5	9
5	4	9	2	1	3	6	8	7

Puzzle 85

2	3	6	7	9	5	1	8	4
9	5	4	6	8	1	3	2	7
8	7	1	4	3	2	9	6	5
6	9	8	3	4	7	2	5	1
5	4	2	9	1	8	6	7	3
7	1	3	2	5	6	4	9	8
3	2	7	5	6	4	8	1	9
4	8	5	1	2	9	7	3	6
1	6	9	8	7	3	5	4	2

Puzzle 86

7	1	3	6	9	2	5	4	8
2	5	9	4	7	8	6	3	1
4	8	6	3	1	5	7	2	9
1	2	5	7	3	9	8	6	4
8	9	7	5	4	6	3	1	2
3	6	4	2	8	1	9	5	7
9	4	2	8	5	3	1	7	6
6	3	1	9	2	7	4	8	5
5	7	8	1	6	4	2	9	3

Puzzle 87

1	9	8	4	6	2	3	7	5
4	5	2	3	1	7	8	9	6
6	3	7	9	5	8	1	4	2
8	2	6	1	4	9	7	5	3
3	1	4	7	8	5	2	6	9
9	7	5	6	2	3	4	8	1
2	4	3	8	9	6	5	1	7
7	6	1	5	3	4	9	2	8
5	8	9	2	7	1	6	3	4

Puzzle 88

3	2	7	5	4	1	8	9	6
5	4	9	8	3	6	1	2	7
6	1	8	9	7	2	3	5	4
4	6	2	1	9	5	7	3	8
8	5	3	4	2	7	9	6	1
7	9	1	3	6	8	2	4	5
1	7	4	2	5	3	6	8	9
2	8	5	6	1	9	4	7	3
9	3	6	7	8	4	5	1	2

Puzzle 89

7	4	9	8	6	1	2	5	3
8	3	1	4	2	5	9	6	7
2	6	5	9	7	3	1	4	8
1	5	2	6	3	7	4	8	9
4	9	6	1	5	8	7	3	2
3	7	8	2	4	9	6	1	5
6	2	3	5	9	4	8	7	1
9	8	7	3	1	6	5	2	4
5	1	4	7	8	2	3	9	6

Puzzle 90

3	8	9	6	1	7	2	4	5
2	5	7	3	4	8	1	9	6
6	1	4	2	5	9	8	3	7
9	3	1	8	6	2	7	5	4
4	7	8	5	9	1	6	2	3
5	2	6	4	7	3	9	8	1
7	9	5	1	8	4	3	6	2
8	6	3	7	2	5	4	1	9
1	4	2	9	3	6	5	7	8

Puzzle 91

5	3	8	6	7	2	9	4	1
4	9	1	5	8	3	2	7	6
6	7	2	4	9	1	8	5	3
1	2	5	7	3	9	6	8	4
7	8	4	2	6	5	1	3	9
9	6	3	8	1	4	7	2	5
2	1	6	3	4	8	5	9	7
8	4	9	1	5	7	3	6	2
3	5	7	9	2	6	4	1	8

Puzzle 92

9	5	2	8	7	3	1	4	6
6	1	8	5	4	9	3	2	7
4	3	7	1	2	6	9	8	5
2	7	3	4	1	5	8	6	9
1	6	5	9	8	7	4	3	2
8	4	9	6	3	2	5	7	1
5	8	4	2	6	1	7	9	3
7	2	1	3	9	8	6	5	4
3	9	6	7	5	4	2	1	8

Puzzle 93

1	4	3	5	2	7	9	6	8
2	6	8	9	3	4	5	1	7
7	9	5	8	1	6	2	3	4
4	5	2	6	9	8	1	7	3
3	1	6	2	7	5	8	4	9
8	7	9	1	4	3	6	5	2
5	8	7	4	6	9	3	2	1
6	3	1	7	8	2	4	9	5
9	2	4	3	5	1	7	8	6

Puzzle 94

6	9	2	1	4	7	3	8	5
5	4	8	2	3	9	7	6	1
1	7	3	8	6	5	2	9	4
7	3	9	4	5	1	6	2	8
4	2	5	6	8	3	1	7	9
8	6	1	9	7	2	5	4	3
9	5	6	7	1	8	4	3	2
3	8	4	5	2	6	9	1	7
2	1	7	3	9	4	8	5	6

Puzzle 95

6	4	7	2	8	1	5	3	9
3	1	2	9	6	5	8	7	4
9	5	8	3	4	7	1	2	6
7	2	3	6	9	8	4	5	1
1	8	6	4	5	2	7	9	3
4	9	5	1	7	3	6	8	2
2	7	1	8	3	6	9	4	5
8	6	9	5	2	4	3	1	7
5	3	4	7	1	9	2	6	8

Puzzle 96

1	5	6	7	2	4	9	8	3
8	3	9	6	5	1	4	7	2
2	4	7	8	3	9	6	1	5
6	1	3	2	7	5	8	9	4
7	8	5	4	9	3	2	6	1
9	2	4	1	6	8	3	5	7
4	9	1	5	8	2	7	3	6
5	7	8	3	4	6	1	2	9
3	6	2	9	1	7	5	4	8

Puzzle 97

8	5	9	4	2	7	3	1	6
4	2	7	1	6	3	9	8	5
3	6	1	8	9	5	2	7	4
9	1	2	7	3	6	5	4	8
7	3	4	9	5	8	6	2	1
5	8	6	2	4	1	7	3	9
6	7	8	5	1	2	4	9	3
2	4	3	6	8	9	1	5	7
1	9	5	3	7	4	8	6	2

Puzzle 98

3	5	1	7	8	9	4	6	2
2	6	9	5	4	3	1	8	7
8	7	4	2	6	1	3	9	5
7	3	8	1	9	2	6	5	4
1	2	5	6	3	4	8	7	9
9	4	6	8	5	7	2	3	1
5	1	3	4	7	6	9	2	8
6	8	2	9	1	5	7	4	3
4	9	7	3	2	8	5	1	6

Puzzle 99

2	6	7	9	8	3	4	1	5
1	9	8	5	4	2	7	3	6
5	4	3	1	6	7	2	9	8
8	3	4	2	5	6	1	7	9
9	7	5	8	3	1	6	4	2
6	2	1	7	9	4	8	5	3
7	5	6	4	2	9	3	8	1
3	1	9	6	7	8	5	2	4
4	8	2	3	1	5	9	6	7

Puzzle 100

9	6	1	2	7	8	4	3	5
7	3	2	4	5	9	1	8	6
5	4	8	1	3	6	7	9	2
8	2	6	7	9	4	5	1	3
4	1	9	3	6	5	8	2	7
3	7	5	8	1	2	9	6	4
2	8	3	9	4	7	6	5	1
1	5	4	6	8	3	2	7	9
6	9	7	5	2	1	3	4	8